A Year on a Farm

Nancy Dickmann

BROWN BEAR BOOKS

Published by Brown Bear Books Ltd
4877 N. Circulo Bujia, Tucson, AZ 85718, USA
and
16 Sun Street, Isleham, CB7 5RT, UK

ISBN 978-1-83572-128-5 (ALB)
ISBN 978-1-83572-134-6 (ebook)

Library of Congress Cataloging-in-Publication Data available on request

Design Manager: Keith Davis
Children's Publisher: Anne O'Daly
Picture Manager: Sophie Mortimer

Manufactured in the United States of America
CPSIA compliance information: Batch#AG/5670

Picture Credits
Front Cover: Shutterstock: Africa Studio tl, arrogant cl, Eric Isselée br, JeniFoto tr, Nataliya Schmidt bl: **Interior: Shutterstock:** Alchemist from India 10, Alter-ego 1, Begir 12, Max Belchenko 16, Alessandro Biascioli 4, Marina Boiko 17, Brick House Farms 19, Callipso88 14, Coatesy 8, domnitsky 23, Foto Kostic 5, 7, Galdrific 20, Eric Isselée 3, Irina Kononova 9, Vladimir Kovrizhnik 22bl, Aleksandar Malivuk 22t, Luce Morin 6, 15, Alioja Neumiler 22br, Milan Rybar 18, Attasit Saentep 21, Matthew J. Thomas 13, Tunatura 11.
t=top, b=bottom, l=left, r=right, c=center
All other graphics and backgrounds Shutterstock.

Brown Bear Books has made every attempt to contact the copyright holder. If you have any information about omissions, please contact: licensing@brownbearbooks.co.uk

Websites
The website addresses in this book were valid at the time of going to press. However, it is possible that contents or addresses may change following publication of this book. No responsibility for any such changes can be accepted by the author or the publisher. Readers should be supervised when they access the Internet.

Words in **bold** appear in the Words to Know on page 23.

Contents

Seasons on a Farm.................... 4
Plowing and Planting 6
Baby Animals.............................. 8
Growing Plants 10
Keeping Cool............................. 12
Storing Food.............................. 14
Harvest Time 16
Staying Warm............................ 18
Fixing and Planning................. 20
Activity......................................22
Words to Know 23
Find Out More 24
Index 24

Seasons on a Farm

On a farm, farmers grow food for us to eat. They grow **crops** and raise animals. Farmers depend on the weather. In many places, the weather is different each **season**. Farmers are busy all year round.

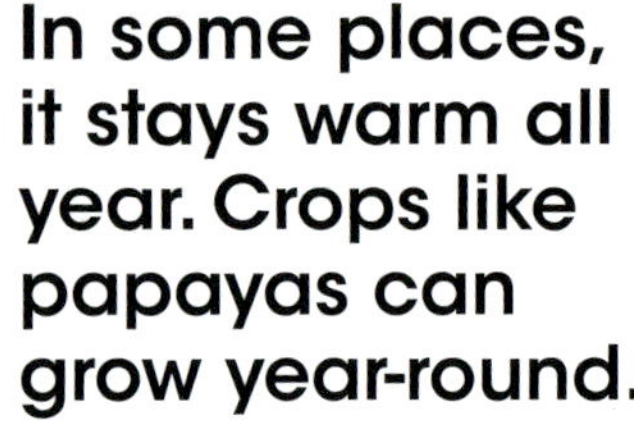

In some places, it stays warm all year. Crops like papayas can grow year-round.

Being a farmer is an important job! There is a lot to do. Farmers do different jobs in spring, summer, fall, and winter. A lot changes on a farm as the seasons change.

Plowing and Planting

In spring, it's time to plant. Farmers get their fields ready. They **plow** the fields to turn over the soil. A strong tractor pulls the plow. Then farmers use another tool to break up the soil.

A plow turns over large chunks of soil.

Farmers plant many thousands of seeds. The seeds will grow into plants. Machines help sow the seeds. Some farmers plant wheat or corn. Others plant carrots or cabbages.

Baby Animals

Spring is a time of new life. Many animals have babies in the spring. Baby lambs are born at the start of spring. Farmers help the mother sheep if needed. They check over the baby lambs.

Mother sheep usually have one or two babies each.

Cows often have babies in the spring, too. At this time of year, there is plenty of grass to eat. Mother sheep and cows feed their babies milk. The babies soon start to eat grass as well.

Growing Plants

In summer, plants grow tall and strong. They need water and sunlight to grow well. Farmers keep a close eye on their crops. They check that plants are growing well. If the weather is dry, they water them.

Farmers put **fertilizer** on their crops. It's like vitamins to help plants grow.

Bugs and other **pests** like to eat farm crops. Some farmers spray chemicals to kill pests. Others use natural methods. They bring in other animals, such as ladybugs. These animals eat the pests.

Keeping Cool

It can be very hot in the summer. Farmers help their animals stay cool. They make sure they have shade. Some farmers put up fans or water sprays to help. They feed the animals at cooler times of day.

Farmers give their animals plenty of water in summer.

Some animals get wet to cool down. Pigs can't sweat, so they roll in cool mud. When the mud dries, it cools the pigs. Other animals lie flat on the ground. This helps cool their bodies.

Storing Food

Many farm animals **graze** on grass. In the winter, there is less grass. Farmers feed them hay instead. Hay is dried grass. Farmers grow grass when it is warm. Then they dry it and store it for winter.

Horses graze on grass in the summer. So do cows and sheep.

In the summer, farmers cut grass for hay.
The grass lies in the field to dry out.
A machine rolls the hay into big bales.
The bales go into a barn, ready for winter.

Harvest Time

In late summer, many crops are ready to eat. Farmers pick fruits and vegetables. They **harvest** grains like wheat and corn. Each crop is ready at a different time. Farmers choose the best time to harvest.

High-tech tools help farmers. They tell them when crops are ready.

WOW!

Some crops get planted in the fall! Farmers can plant winter wheat. It will grow over the winter.

As summer turns into fall, days get shorter. Temperatures are cooler, and leaves change color. Crops like carrots and pumpkins are ready now. So are apples and potatoes. Farmers pick them before winter comes.

Staying Warm

Winter can be very cold. Farm animals need to stay warm. Some, like cows and sheep, grow thicker fur. They huddle together in groups. This helps keep them warmer. Some animals move indoors.

Thick fur traps an animal's body heat. It keeps them warm in cold weather.

A coop provides chickens with shelter from the cold. Cows often move into a barn. Farmers feed them the hay that they made in the summer. They put down thick, dry straw. It makes a warm, cozy bed.

Fixing and Planning

In the winter, farmers spend less time working in the fields. Some even take a short vacation! But it's still a busy time. Farmers see what needs doing around the farm. They mend fences and fix buildings.

Some farmers use the winter as a time to fix machines.

WOW!

Some jobs must be done in the winter. This is when farmers trim apple trees. Trimming helps them grow better.

Winter is also a time to plan for spring. Farmers decide what to plant. They buy seeds and fertilizer. They might even learn new skills. Soon the next farming year will begin. They will be ready!

Activity

Here are some jobs on a farm. Match each one to the season when it is done. The answers are on page 24.

1. Farmers do this job in the spring.
2. Farmers do this job in the fall.
3. Farmers do this job in the winter.

Words to Know

crops plants that are grown for food

fertilizer a substance added to soil to make it richer and better for growing plants

graze to eat grass and other plants on the ground in a field

harvest to pick a crop when it is ripe and ready to eat

litter a group of babies that a mother animal gives birth to at the same time

pest an animal that harms crops or annoys farm animals

plow to break up the soil for planting, using a tool with blades, called a plow

season part of the year that has its own temperature and weather patterns

Find Out More

Websites

ffdt.co.uk/explore/the-farming-year

kids.britannica.com/kids/article/agriculture/352715

seeds.ca/schoolfoodgardens/what-do-farmers-do-during-the-winter/

Books

A Trip to the Farm: Crop Farms Ursula Pang, Rosen Publishing, 2023

Farm Animals Bonnie Hinman, Abdo Reference 2023

It's Autumn Ruth Owen, Ruby Tuesday Books, 2024

Index

baby animals 8, 9

chickens 19
cows 9, 14, 15, 18, 19
crops 4, 10, 11, 16, 17

fall 5, 17
fertilizer 10, 21
fields 6, 15, 20

grass 9, 14, 15

hay 14, 15, 19
horses 14

machines 7, 15, 20

picking crops 16, 17
pigs 9, 13
planting seeds 6, 7

sheep 8, 9, 14, 18
spring 5, 6, 8, 9, 21
summer 5, 10, 12, 14, 15, 16, 17, 19

water 10, 12
weather 4, 5, 18
winter 5, 14, 15, 17, 18, 20, 21

Answers: 1. B; 2. C; 3. A